Marion Jr. High Library
2 Patriot Dr.
Marion, AR 72364

EARTH EXPLORATIONS

Rocks and Minerals

Jenny Karpelenia

Editorial Director: Susan C. Thies
Editor: Mary L. Bush
Design Director: Randy Messer
Book Design: Tobi S. Cunningham
Cover Design: Michael A. Aspengren

A special thanks to the following for their
scientific review of the book:

Kristin Mandsager
Instructor of Physics and Astronomy
North Iowa Area Community College

Jeffrey Bush
Field Engineer
Vessco, Inc.

Image Credits:
© Sukree Sukplang/Reuters Newmedia Inc/CORBIS: p. 12 (bottom); © Laura Dwight/CORBIS: p. 13 (top); © José Manuel Sanchis Calvete/CORBIS: pp. 16, 21; © Layne Kennedy/CORBIS: p. 24 (middle); © Roger Ressmeyer/CORBIS: p. 27 (bottom); © Ric Ergenbright/CORBIS: p. 28 (bottom); © Gary Braasch/CORBIS: p. 29 (top); © CORBIS: p. 38

Photos.com: cover, pp. 1, 3, 4, 5 (top), 6, 8, 11, 12 (top), 13 (bottom), 15, 18, 19, 23, 24 (top and bottom), 25, 26, 28 (top), 30, 31, 32, 34, 36, 37, 39, 40, 41, 42, 43, 47; NOAA: p. 20 (top), 29 (bottom); NASA: p. 44; PLC images: pp. 5, 7, 9, 10, 16, 17, 20 (bottom), 21, 22, 33, 35

Text © 2005 by Perfection Learning® Corporation.
All rights reserved. No part of this book may be reproduced, stored in a retrieval system, or transmitted in any form or by any means, electronic, mechanical, photocopying, recording, or otherwise, without prior permission of the publisher. Printed in the United States of America.

For information, contact
Perfection Learning® Corporation
1000 North Second Avenue, P.O. Box 500
Logan, Iowa 51546-0500.
Phone: 1-800-831-4190
Fax: 1-800-543-2745
perfectionlearning.com

1 2 3 4 5 6 PP 09 08 07 06 05 04
ISBN 0-7891-6218-0

Table of Contents

1. A Guide to Geology 4
2. The Third Rock from the Sun 7
3. Minerals 11
4. Crystals 19
5. Meet Some Minerals 23
6. Rocks 27
7. Recycling Rocks 33
8. Mining 37
9. A World of Natural Wealth 41
   Internet Connections and Related Reading
   for Rocks and Minerals 45
   Glossary 46
   Index 48

1

A Guide to Geology

Imagine walking along a lakeshore. The hot Sun warms your face. The sand feels soft between your toes. Bending down, you pick up a smooth, flat **rock**. It's a perfect rock-skipping stone. Flicking your wrist, you hurl the rock so it skims the water's surface. The rock skips across the water. One, two, three, four, five . . . You watch as it sinks into the depths of the lake.

As you continue your walk, you notice some interesting rocks along the shore. One is sparkly. Another one has streaks of different colors running through it. Some are smooth. Others are rough and chunky. You decide to take some home and start a rock collection.

The sand you walked on, the stone you skipped, and the rocks you collected are all types of rock. What are rocks? How are they made? What makes one so different from another?

Geology is the study of these questions and more. Geologists study the Earth and its features. They map the land as it is today and compare it to the past. They examine **elements**, **minerals**, rocks, soil, and landforms to help determine how the Earth has changed over time.

Geologists study for many years to prepare them for their exploration of the Earth. Three important words for geologists are *elements*, *minerals*, and *rocks*.

Elements

Elements are nonliving materials made up of one kind of **atom**. All matter, or stuff, in the world is made up of elements. There are more than 100 elements known in the world today. Oxygen, hydrogen, helium, carbon, iron, gold, and silver are some common elements.

The Periodic Table

The elements are arranged on a chart called the *periodic table*. This chart groups the elements according to common characteristics.

1 H																	2 He
3 Li	4 Be											5 B	6 C	7 N	8 O	9 F	10 Ne
11 Na	12 Mg											13 Al	14 Si	15 P	16 S	17 Cl	18 Ar
19 K	20 Ca	21 Sc	22 Ti	23 V	24 Cr	25 Mn	26 Fe	27 Co	28 Ni	29 Cu	30 Zn	31 Ga	32 Ge	33 As	34 Se	35 Br	36 Kr
37 Rb	38 Sr	39 Y	40 Zr	41 Nb	42 Mo	43 Tc	44 Ru	45 Rh	46 Pd	47 Ag	48 Cd	49 In	50 Sn	51 Sb	52 Te	53 I	54 Xe
55 Cs	56 Ba	57 La	72 Hf	73 Ta	74 W	75 Re	76 Os	77 Ir	78 Pt	79 Au	80 Hg	81 Tl	82 Pb	83 Bi	84 Po	85 At	86 Rn
87 Fr	88 Ra	89 Ac	104 Rf	105 Db	106 Sg	107 Bh	108 Hs	109 Mt	110 Uun	111 Uuu	112 Uub						

58 Ce	59 Pr	60 Nd	61 Pm	62 Sm	63 Eu	64 Gd	65 Tb	66 Dy	67 Ho	68 Er	69 Tm	70 Yb	71 Lu
90 Th	91 Pa	92 U	93 Np	94 Pu	95 Am	96 Cm	97 Bk	98 Cf	99 Es	100 Fm	101 Md	102 No	103 Lr

Minerals

Minerals are nonliving substances made up of elements. Copper and lead are elements that are also minerals. Quartz is a mineral that is a combination of the elements silicon and oxygen. There are more than 3000 minerals on Earth.

Copper pipes and tubing

Rocks

Rocks are combinations of two or more minerals. Rocks form in different ways and change over time. These "chunks" of the Earth have fascinated people throughout history.

2

The Third Rock from the Sun

The Earth is a gigantic ball of minerals and rock. It is often called the "third rock from the Sun" because it is the third-closest planet to the Sun. Earth is made up of three layers—the **core**, the **mantle**, and the **crust**.

The Core

The core is the center of the Earth. It is divided into the inner and outer core. The inner core is a solid layer of metal made of mostly iron and nickel. It is located more than 3000 miles below the Earth's surface. The radius (distance from the center to the outer edge) of the inner core is about 800 miles. The inner core reaches a temperature of more than 10,000°F.

A Message About Metals

Metals are minerals that have special characteristics, such as a shiny surface, a weight greater than water, the ability to **conduct** electricity and heat, and the flexibility to be made into sheets or wires. Not all metals have all of these characteristics, but they do share a mixture of these traits. Silver, gold, copper, tin, and iron are examples of metals.

The outer core is molten, or liquid, rock. This layer is about 1400 miles wide and approximately 7000°F. Most of the melted rock is iron and nickel. These flowing metals make the Earth act as a huge magnet.

The Mantle

The layer above the core is the mantle. This layer is about 1800 miles wide. The temperature in the mantle varies from about 1600°F at the top to 5000°F at the bottom. This hot rock is a soft solid in many areas. Some places in the mantle, however, get so hot that the rock melts and becomes a thick liquid. Hotter areas flow upward, cool, and sink down again. These slow currents in the mantle are what cause the **plates** in the Earth's crust to move.

Planet Peach

To help you remember the layers of the Earth, imagine a peach cut in half lengthwise. Its layers are like those of the Earth. The peach skin is like the Earth's crust. The fruit is the mantle. The peach pit is the core.

The Crust

The outer layer of the planet is the crust. The crust is about 25 to 40 miles thick beneath the continents. Under the oceans, the crust is thinner, averaging about 4 miles deep. Most of the crust is made of the elements silicon, aluminum, calcium, oxygen, sodium, and potassium.

Plates

The Earth's crust is divided into big pieces called *plates*. These plates move very slowly on top of the flowing mantle. When plates push against one another and push rock upward, mountains are formed. Colliding plates can also cause volcanoes and deep sea trenches. When plates move apart, volcanoes and ocean ridges can form. Earthquakes occur when plates slip past one another.

Plate Science

The scientific study of the movement of the Earth's plates is called *plate tectonics*.

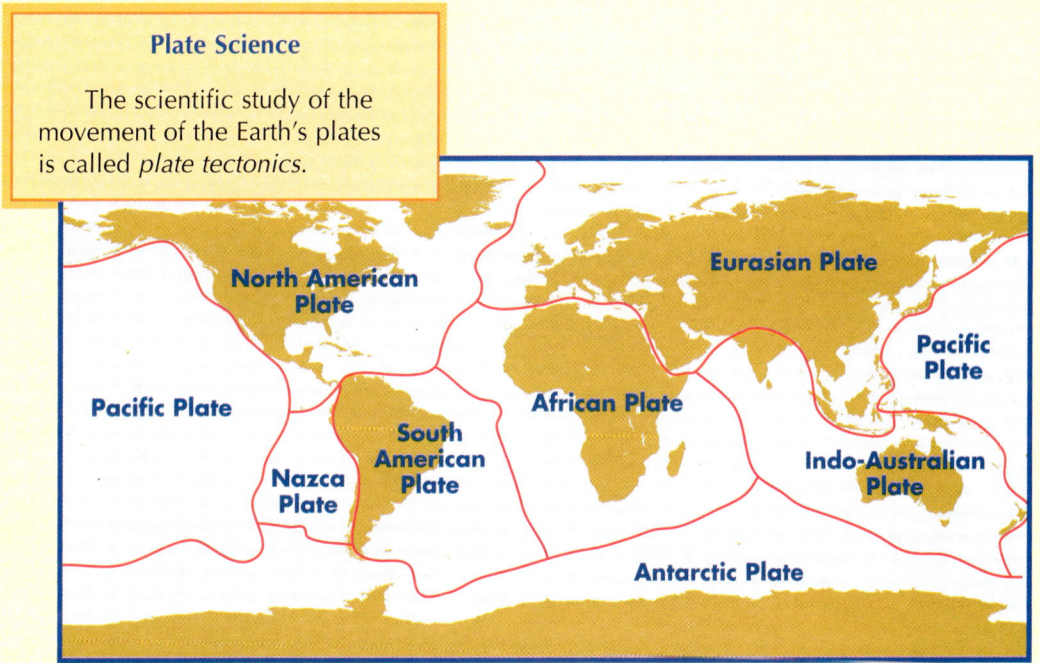

TRY THIS!

Make an edible Earth model. Open a pudding cup of any flavor. The pudding is the hot, molten mantle. Press a piece of soft taffy around a gumball. Push the ball down into the center of the pudding. The taffy is the outer core and the gumball is the inner core. Place a vanilla wafer on top of the pudding to form the crust. If you'd like, spoon a dab of frosting on top of the wafer to form a mountain. Top the mountain with colored sprinkles or sugars or any other small candy to add rocks. Then enjoy eating your edible Earth!

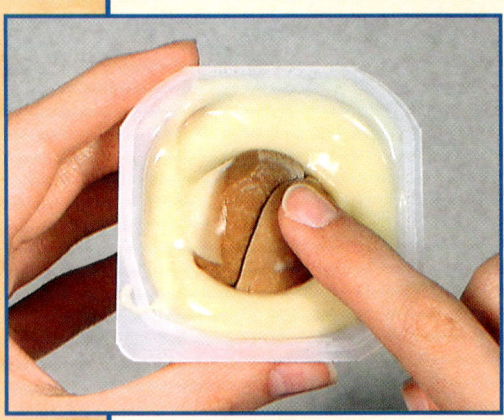

Minerals

Minerals are solid materials made up of one or more elements. There are more than 3000 different minerals.

Minerals are classified by their particular properties, or characteristics. The way the atoms in a mineral are arranged gives the mineral its properties. The most common properties used to identify a mineral are color, luster, streak, hardness, cleavage, habit, and specific gravity.

Millions of Pounds of Minerals

Do you know how many pounds of copper, aluminum, or coal you will use over your lifetime? The average American needs 1935 pounds of copper, 5929 pounds of aluminum, and 589,974 pounds of coal throughout his or her life. All together, a person uses 3.75 million pounds of minerals, metals, and fuels during a lifetime!

Copper

Color

The color of a mineral can be an identifying property. Some minerals only exist in one color. Sulfur, for example, is always yellow. The mineral gold also has a characteristic color.

Amethyst

Other minerals exist in a variety of colors. Chemical impurities within a mineral can change its color. For example, the mineral quartz is white or colorless when pure. But if iron is mixed in the quartz as **crystals** are forming, it will become purple in color. Purple quartz is called *amethyst*.

This 200-karat blue sapphire has an estimated worth of $1.2 million.

Corundum is another mineral that is colorless when pure. When the element chromium colors it red, it becomes a ruby. Corundum is called a *sapphire* when the elements iron and titanium color it blue.

Luster

Luster is another word for "shine." A mineral's ability to reflect light helps determine its luster. Words used to describe a mineral's luster include *shiny, bright, metallic, glassy, greasy, pearly, silky, waxy,* and *dull*. Quartz has a glassy luster. Copper, gold, and silver have a metallic luster. Talc, which is used to make talcum powder (baby powder), has a pearly luster.

Streak

A mineral's streak is the color that it makes when it's crushed. When a mineral is rubbed against a white unglazed ceramic tile, some of the mineral is crushed into a powder. This leaves a streak of color on the tile.

Sometimes the mineral's streak color is the same as its outside color. An example of this is the mineral lazurite. Lazurite is a blue mineral that leaves a blue streak.

Sometimes a mineral's streak color is different from its outside color. The mineral jade is usually green in color, but it leaves a white streak when crushed.

Jade is actually the name given to two different minerals—jadeite and nephrite.

Hardness

Did you know that not all minerals are hard? Some are actually very soft. A mineral's hardness is measured by its ability to withstand being scratched and its ability to scratch other minerals.

In 1812, Friedrich Mohs, a mineralogist (mineral expert) from Austria, devised a hardness scale that is still used today. Mohs chose certain minerals and assigned them numbers on the scale based on how hard they were compared to one another. The harder the mineral, the higher its number. Any mineral with a lower number can be scratched by a mineral with a larger number. All minerals fall somewhere on this scale of hardness.

Mohs Scale of Hardness

Hardness Number	Mineral Example	Item(s) That Will Scratch Minerals of This Hardness
1	talc	fingernail (easily)
2	gypsum	fingernail (Fingernails have a hardness of 2.5.)
3	calcite	copper penny (A penny has a hardness of 3.5.)
4	fluorite	pocket knife, common nail
5	apatite	glass
6	orthoclase	concrete nail
7	quartz	steel file (Steel has a hardness of 7–7.5.)
8	topaz	sapphire, ruby
9	corundum	diamond (four times as hard as corundum)
10	diamond	none

Hardness can be used to identify two similar minerals. For example, the mineral gold is very soft. It has a hardness of 2.5–3 on the scale. A mineral often confused with gold, called *pyrite* or *fool's gold*, is much harder. It has a hardness of 6–6.5 on the scale. This difference in hardness can be used to tell these two minerals apart.

TRY THIS!

Gather ten common items made of minerals (pencil lead, chalk, glass, penny, etc.). Choose an item and use it to scratch the others. Record the results. Rotate through the items using each one as the "scratcher." Using your results, put your items in order of hardness from softest to hardest.

Cleavage

Cleavage refers to how a mineral breaks. There are weak areas in some minerals called *cleavage planes*. Cleavage planes occur because of the way a mineral's atoms are arranged. The bonds holding the atoms together are stronger in parts of the pattern but weaker in others. When a mineral is struck, it breaks along its cleavage planes. A mineral with perfect cleavage always breaks in the same direction.

There are several types of cleavage. Each type is characterized by the shape, direction(s), and angle(s) of the break. Some minerals, like muscovite and biotite, break in one direction into sheets. Prism cleavage, such as that found in the mineral amphibole, results in breaks in two directions. The mineral feldspar has block cleavage, which means it breaks in two directions at right angles. Rhombic cleavage occurs in minerals, such as calcite, with breaks in three directions. Galena and halite are minerals that break in three directions at right angles, which is called *cubic cleavage*. Fluorite is a mineral that breaks in four directions.

> **A Prism Picture**
>
> A prism is a columnlike structure with at least two parallel sides.

Types of Cleavage

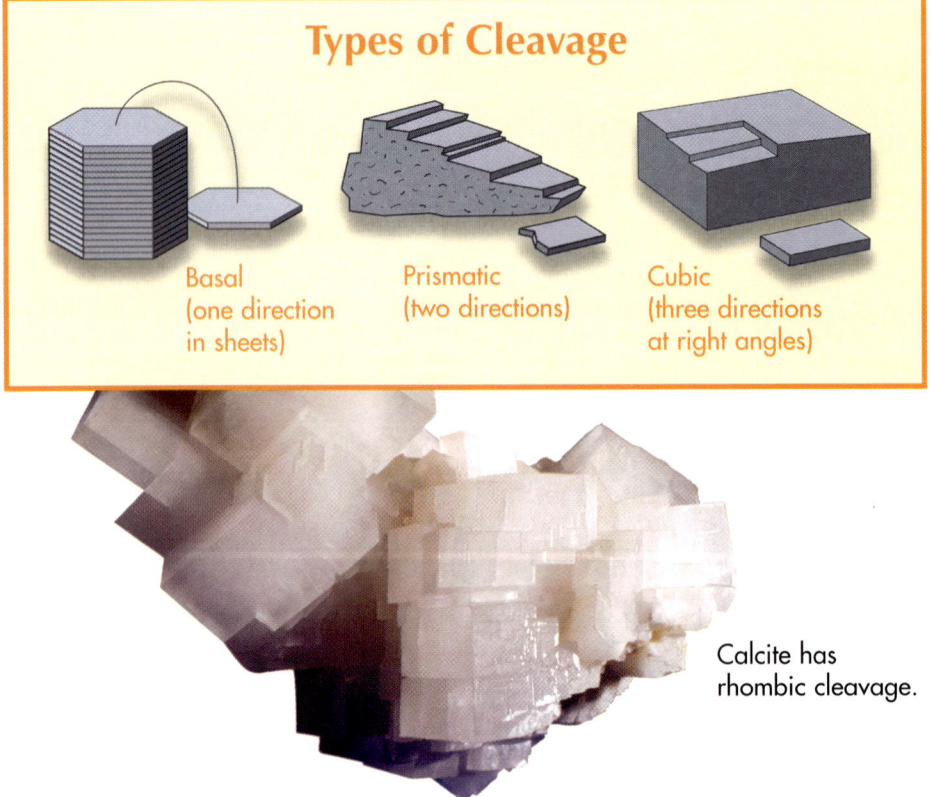

Basal (one direction in sheets)

Prismatic (two directions)

Cubic (three directions at right angles)

Calcite has rhombic cleavage.

TRY THIS!

Gather two sheets of paper towel. Rip one sheet in half in one direction. Examine the tear line. How hard was it to tear? Now rip the other sheet in half in the other direction. Compare the two tears and the effort it took to rip the towels. What do you notice?

You should notice that the sheet tears much straighter and easier in one direction. The atoms in paper towels have weak areas just like the atoms in minerals. The sheet can be torn more easily along the weaker lines.

Habit

Minerals are found in different shapes. The shape a mineral forms in nature is called its *habit*. Just like *your* habits are things you do on a regular basis, a *mineral's* habit is the form it usually takes when left alone to grow. Interfering with a mineral's growth by moving it, breaking it, squeezing it into a smaller space, etc., can cause the mineral to take a shape other than its habit.

There are many types of habits. The chart below shows some common habits.

Type of Habit	Shape of the Mineral	Mineral Example
nugget	chunks or lumps	gold
radial	similar to spokes on a bike wheel	pyrophyllite
dendritic	similar to tree branches	copper
acicular	needlelike	rutile
prismatic	prism	quartz
reniform	similar to kidneys	hematite
rosette	similar to flower petals	gypsum

Habit Help

To find a complete list of habits, check out these Web sites.
http://webmineral.com/help/Habits.shtml
http://web.wt.net/~daba/Mineral/tablehabit.html

Specific Gravity

Another property used to identify minerals is specific gravity. This is a measurement of how heavy the mineral is compared to an equal amount of water. Quartz has a specific gravity of 2.6. This means it weighs 2.6 times as much as an equal amount of water. With a specific gravity of 8.9, nickel weighs almost 9 times as much as water. Silver and gold are extremely heavy metals with specific gravities of 10.5 (silver) and 19.3 (gold).

Gold bars

Crystals

Most minerals take the form of crystals. Crystals are solids that grow with their atoms arranged in a regular, repeated pattern. The smooth, flat sides (faces) of crystals can be shaped like squares, rectangles, triangles, or hexagons.

Crystals can be found in many different places. When weaker rock around crystals breaks down and is worn away, the stronger crystals are left behind. Crystals can also be brought to the surface of the Earth by a volcanic eruption. Riverbeds are another source of crystals. The crystals are washed downstream with the river and then **deposited** with **sediment** like sand and gravel.

Crystal Formation

Crystals form in several ways. Often they form when the water they are mixed with **evaporates**, leaving the mineral behind. This is called *precipitating*. When a volcano erupts, hot **lava** overflows. As the liquid rock cools, minerals crystallize. Crystals can also form when liquids freeze and gases change into a solid form.

Many crystals grow as pairs called *twins*. Twins can grow on opposite sides of each other. One twin can even grow through the other.

Lava flow

It usually takes a long time for crystals to form. The longer the time and the more room they have to grow, the bigger the crystals get.

TRY THIS!

To get an idea of how crystals form, try growing these quick crystals. Fill a glass jar half full with water. Add several teaspoons of salt or sugar and stir until mixed. Tie a piece of string around a pencil. Lay the pencil across the top of the jar so the string hangs in the water. The string should be long enough so that an inch or so of it lies on the bottom of the jar. Watch your crystals form as the water evaporates from the jar. (This will take several days or weeks depending on the size of the jar/amount of water.)

Crystal Shapes

There are many crystal shapes, but they are usually grouped into six basic shapes. The groups are based on the number of faces, the angles between the sides, and the length of the sides.

Six Crystal Shape Groups

Group	Shape	Mineral Examples
isometric	faces that are squares and triangles	halite, galena, gold, pyrite
tetragonal	four-sided prism or two pyramids stuck together at their bases	zircon
hexagonal	six-sided prism or pyramid	quartz, calcite
orthorhombic	short and stubby	sulfur, topaz, olivine
monoclinic	short and stubby with ends at tilted angles	gypsum, mica, talc
triclinic	flat with sharp edges	feldspar, turquoise

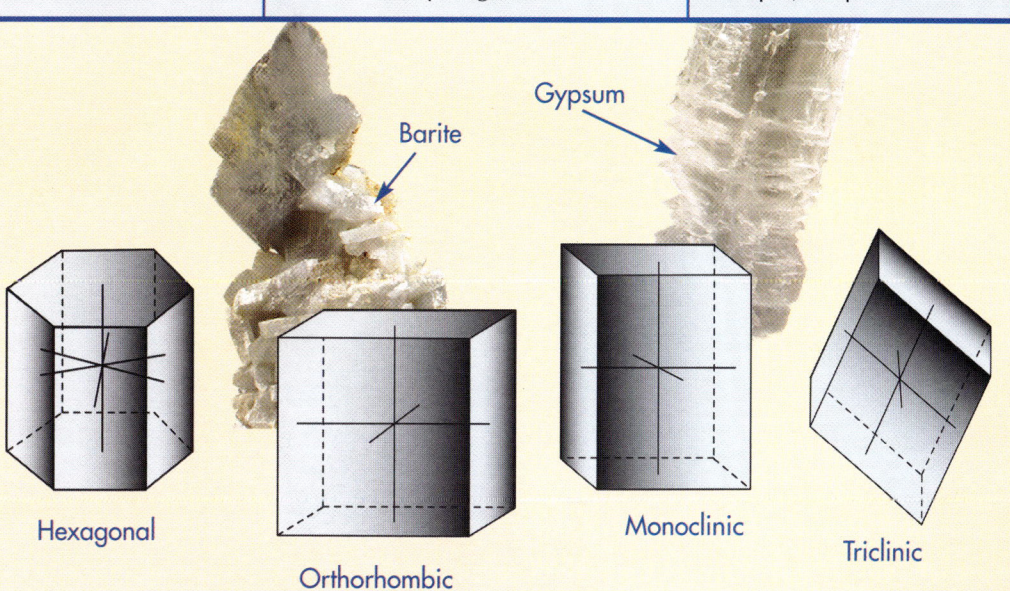

TRY THIS!

Gather some uncooked spaghetti noodles and small marshmallows. Use the noodles to represent straight lines and the marshmallows to form corners. Make a model of each of the six basic crystal shapes.

5

Meet Some Minerals

A mineralogist has more than 3000 minerals to learn and study. In everyday life, most people are only familiar with the more common minerals. Let's examine of a few of these.

Quartz

Quartz is made of the elements silicon and oxygen. Its color is often clear, but it can also be found as purple amethyst, yellow citrine, or rose quartz. Quartz is a strong mineral. It forms when hot water carrying the mineral flows into the cracks and holes in rocks. Here, it cools and forms crystals. Special rocks called *agates* can be split open to reveal many beautiful quartz crystals growing inside. Quartz is also found in sand. Glass and computer microchips are made with quartz.

Quartz

What a Gem!

Has anyone ever told you that you're a "gem"? Do you know what a gem is? In rock terms, a gemstone is a rock or mineral used for decoration, such as in jewelry. Gemstones are rare, beautiful, and strong. Diamonds, rubies, emeralds, amethysts, and garnets are gemstones.

So, if someone calls you a gem, he or she means that you are precious like a gemstone.

Pearl

Pearls are made with the help of oysters and other shelled water animals. If a tiny bit of sand or other particle becomes stuck inside the animal's shell, the animal covers it with a substance called *nacre*. Layer after layer of nacre surrounds the grain over time, forming a pearl. Pearls are softer gemstones used for jewelry.

> **Sticky Stuff**
>
> Amber is an interesting mineral formed with the help of a plant. Long ago, amber began as pine tree sap. When the trees were buried underground or underwater, the sap hardened into a mineral. Insects were often trapped inside the flowing sap and became a part of the hardening mineral.

Diamonds

Diamonds are forever. Diamonds are a girl's best friend. He's a diamond in the rough. Have you ever heard any of these expressions?

Diamonds are a very popular mineral. These gemstones are formed in the Earth's upper mantle. Because of the great pressure at this depth, a diamond's atoms are squeezed together tightly. This is why diamonds are the hardest substance known. They are often put on the tips of saws, drills, surgeons' scalpels, and other cutting tools to make them incredibly sharp.

Diamonds look quite dull and rough when found in nature. After they're cut and polished, they sparkle and shine. So "a diamond in the rough" is someone who may look tough on the outside but is kind and gentle on the inside.

Graphite

Did you know that a diamond and your pencil lead are made of the same thing? Both a diamond and lead are made of the element carbon. Pencil lead is actually a mineral called *graphite*. Graphite is a soft, shiny, dark gray mineral with a greasy feel. It is used for writing because it rubs off so easily.

Silver

Silver is a mineral with a shiny, metallic luster. It is harder than gold but can still be hammered into thin sheets. Silver is a good conductor of heat and electricity. North America, including the United States and Canada, is the largest silver producer in the world. Because silver is very soft, it is usually mixed with copper to make other items. This metal mixture, or **alloy**, is used to make silverware, jewelry, and coins.

Sulfur

Sulfur is a very soft yellow mineral. When combined with oxygen, it gives off a strong "rotten egg" smell. Sulfur forms near volcanoes and hot **springs**. It is also found in natural gas and petroleum. Sulfur is used in gunpowder, explosives, and match heads. It is also added to rubber to make it more flexible and less brittle.

Morning Glory hot spring at Yellowstone National Park

Halite

Halite is a mineral you can eat! It is more commonly known as salt. Halite is **mined** from old seabeds. When salty oceans or seas dry up, the salt crystals are left behind. Salt can also be collected from evaporating ocean water.

Solar salt mining is a process that uses the Sun and wind to evaporate salt water out of shallow ponds. The salt crystals are then scraped from the salt bed by a mechanical harvester.

6

Rocks

Minerals are the building blocks of rocks. When one or more minerals combine, a rock is formed. This formation can happen in one of three ways. The three types of rocks that result are **igneous**, **sedimentary**, and **metamorphic**.

Igneous Rocks

Igneous rocks are formed when hot liquid rock from within the Earth cools as it approaches or reaches the Earth's surface. Liquid rock underground is known as **magma**. When magma cools, it forms igneous rocks. There are two types of igneous rocks—intrusive and extrusive.

> **Flaming Rocks**
>
> The word *igneous* means "from fire."

> **Magma or Lava? What's the Difference?**
>
> Magma is hot liquid rock *under* the ground. Lava is hot liquid rock *above* the ground. So lava is magma that has reached the Earth's surface.

Natural lava tubes, or channels, carry magma from the Mount Kilauea Volcano in Hawaii to the Pacific Ocean.

Intrusive

Intrusive igneous rocks are formed when magma does not reach the Earth's surface. Instead, the magma cools within the crust.

Granite is an example of an intrusive rock. Because magma in the crust cools slowly, the crystals have a long time to form. This results in the larger crystals often found in granite. Granite is usually pink or white with large flecks of black and white. The four presidents' faces on Mount Rushmore in South Dakota were blasted and carved out of granite. Polished granite is used as a decorative stone in buildings and countertops.

Mount Rushmore

Extrusive

Rocks that form when magma reaches the Earth's surface are called *extrusive rocks*. Many of these rocks are a result of erupting volcanoes. When lava explodes out of a volcano and meets the air, it cools and hardens quickly. Lava that cools this way has small crystals.

Obsidian is an extrusive igneous rock formed from volcanic lava. Obsidian looks like black glass. It has very smooth surfaces and often has sharp edges.

Obsidian

> **Exploding Extrusives**
>
> Because many extrusive rocks form when lava explodes out of volcanoes, they are also called *volcanic rocks*.

Pumice is a another kind of extrusive igneous rock. Pumice contains many pores, or holes, like a sponge. Lava is very frothy and bubbly. As the lava cools and hardens, gases get trapped in the rock as air bubbles. These pores make the rocks very lightweight. Some can even float.

Some extrusive rocks are formed as magma slowly seeps to the surface of the Earth through cracks in the crust. Basalt is an igneous rock formed by this process. The ocean floor is made of basalt. Magma is pushed up through cracks in the ocean floor. The rock cools quickly as it hits the ocean and solidifies into rock. Basalt is a dark gray or black rock with small crystals.

Pumice on Mount St. Helens Volcano

29

Sedimentary Rocks

Three-fourths of the rocks in the Earth's crust are sedimentary rocks. Sedimentary rocks are formed when layers of particles are pressed together over time. Sediments are small pieces of sand, rocks, shells, or minerals carried by water. When the water deposits sediments, they build up. Over time, pressure squeezes them into sedimentary rock.

A variety of sediments in different colors can result in striped layers of rock. This can be seen in the dark and light layers of limestone, shale, sandstone, and other rock in the Grand Canyon.

Grand Canyon

Shale is an example of sedimentary rock. When small grains of mud and clay are pressed together in layers, shale forms. Shale is a gray rock that is ground up to make bricks and **cement**.

Sandstone is another sedimentary rock. When layers of sand grains are deposited by an ocean or river, the weight of the upper layers pushes down on the lower layers. The water contains minerals that act as a glue. The sand particles are "glued" together, creating sandstone. Because the sand for this rock was left behind by water, scientists know that areas where sandstone is found were once covered with water.

Limestone is another rock that forms where water once existed. Layers of shells from snails, oysters, clams, and other water creatures form this sedimentary rock. When a shelled ocean animal dies, its body sinks to the ocean floor. The soft body of the animal rots, and the hard shell remains. Over time, the weight of upper shell layers pushes down on lower layers. The resulting rock is limestone.

TRY THIS!

Shells are made from a mineral called *calcite*, which contains a chemical called *calcium carbonate*. Rocks that contain calcium carbonate will fizz and bubble when acid is placed on them. The bubbles are carbon dioxide gas. This is one test that can be used to help identify limestone.

Perform your own calcium carbonate test. Gather a piece of chalk, a tablespoon of vinegar, a bowl, and an eyedropper. Put the chalk in the bowl. Squirt a few drops of vinegar on the chalk. What happens? Chalk is made of calcium carbonate, so it will fizz and bubble when acid is added.

If you have a rock that you believe to be limestone, try the test on it.

How Are a Butterfly and a Rock Alike?

Metamorph means "to change shape or form." When a caterpillar builds a cocoon and changes into a butterfly, it is called a *metamorphosis*. Igneous and sedimentary rock that changes due to heat and/or pressure becomes *metamorphic* rock.

Metamorphic Rocks

Metamorphic rocks are igneous or sedimentary rocks that change due to heat and/or pressure within the Earth. The deeper a rock is buried, the greater the heat and pressure. The extremely hot mantle under the crust actually bakes the rock from below. The atoms making up the rock rearrange themselves. This changes the old rock into a new rock with different properties.

One example of a metamorphic rock is marble. Marble begins as limestone. When the limestone is heated and pressed within the Earth, it forms marble. Like limestone, marble will react to acid because it still contains calcium carbonate. However, marble has different qualities than limestone. Marble is a sturdy rock found in a variety of colors, such as red, pink, or green. When polished to a beautiful shine, marble becomes an expensive, precious stone. This pretty rock is used to make pillars, statues, and even floors in buildings.

Gneiss is also a metamorphic rock. When the Earth's crust bends and folds to form mountains, granite underneath the areas is heated and pressed to create gneiss. This metamorphic rock forms large crystals as it cools slowly within the Earth. Gneiss is known for its beautiful bands or stripes of color.

A Nice Rock

The metamorphic rock gneiss is pronounced like the word *nice*.

Slate is a metamorphic rock made from shale. Slate is heavy and breaks into flat plates. Some people have slate roofs. Slate is also used to make chalkboards and pool tables.

Sandstone turns into the metamorphic rock quartzite. Sandstone is quite soft. You can rub off bits of sand or scratch it easily with your fingernail. Water wears away sandstone over time. Quartzite is much stronger than sandstone and comes in a variety of colors. It is often white or gray. Purple quartzite results when the mineral iron is present. Quartzite is crushed and used as railroad beds. Railroad track is then laid down on the beds. This rock is strong enough to support the weight of a train.

Hot and Heavy Rocks

Igneous and metamorphic rock make up more than 90 percent of the rocks found deep beneath the Earth's crust. This is because these rocks are formed by heat and pressure, which increase underground.

7

Recycling Rocks

The Rock Cycle

The three types of rock—igneous, sedimentary, and metamorphic—share an important relationship. They are part of a cycle that "recycles" all the rocks on Earth.

Choose any starting point in the cycle and you can circle all the way around. Let's start at the surface of the Earth. Igneous and metamorphic rocks lie on the Earth's surface. Over time, these rocks are worn down by **weathering** and **erosion**. The sediment that results eventually becomes new sedimentary rock. Changes in temperature and pressure cause sedimentary and igneous rock to change into metamorphic rock. When this metamorphic rock is pushed deep beneath the ground and melts, it becomes magma, which forms new igneous rock. In time, this rock is forced to the surface of the Earth to begin the cycle again.

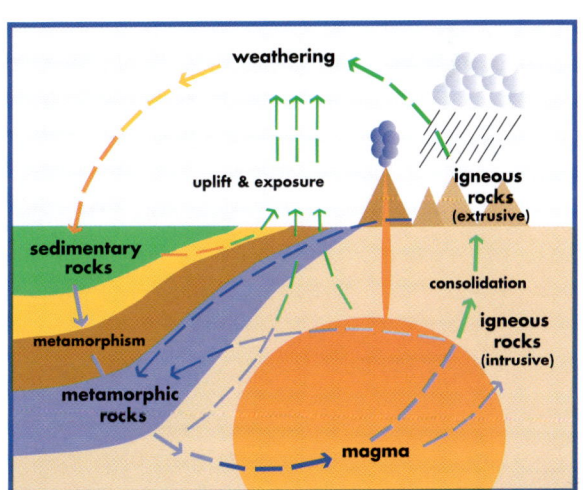

A Closer Look at Weathering and Erosion

What is this weathering and erosion that breaks down igneous and metamorphic rock into sediments? When rock is worn down into smaller pieces, it is called *weathering*. Erosion occurs when these pieces of rock are carried to different places. Wind, water, and ice are the three main forces that cause weathering and erosion. Living things and chemicals also contribute to the process.

Wind

Wind carries tiny sand grains and other particles and pelts them into larger rocks. The grains slowly wear away the rocks' surfaces. Eventually wind can actually carve right through softer rock, leaving natural arches and bridges. Huge amounts of sand moved by wind create sand dunes. This powerful effect of wind erosion is common in deserts.

TRY THIS!

Weathering and eroding rock is much like erasing a pencil mark. Write your name with a pencil on a piece of paper. Then use an eraser to rub off the writing. How is this like weathering and erosion? The eraser wears away the graphite in the pencil lead just as wind, water, or ice scrapes against rock. If you blow away the graphite particles, you're acting like the wind.

Water and Ice

Water is a powerful force that can break down rocks and carry away the particles. Wave action at the shores of oceans and lakes can weather and erode rocks. Rocks caught up in the waves smash into other rocks. This can break the rocks apart. Rocks tumbling around in the water are smoothed into smaller round pebbles. The particles worn off the jagged edges are carried away in the water and end up as sediment.

Rivers and streams can also break down and carry rocks and sediments. Powerful, fast-moving rivers can cut deep V-shaped valleys through rock. The Grand Canyon was carved this way by the Colorado River. The sediments left behind are deposited at the river's banks and mouth (where the river meets an ocean). This sediment builds up and becomes new sedimentary rock.

Rainwater can cause weathering and erosion as well. Heavy downpours or floodwaters can rush across the ground and wear down or carve rock. The Badlands in South Dakota were carved by rainwater.

See how the flow of water affects rock. Place a bar of soap on a sponge in a sink. Run water from the faucet slowly on top of the soap. Check the soap every 5 minutes for 15 minutes. What do you notice?

You should see the water starting to wear a hole in the soap. The soap washed away from the bar was carried down the drain with the water. This action is like the weathering and erosion of rock caused by flowing water.

Rainwater flows into cracks in rocks. When temperatures drop, the water turns to ice. Ice takes up more space than water. It expands and pushes against the rock. It can be so powerful that it breaks the rock apart. This frost action can be seen at Devil's Lake in Wisconsin. The lake is surrounded by huge blocks of purple quartzite that have broken off from the bluffs above.

Glaciers are huge chunks of slow-moving ice. They are immensely heavy and can scrape and carve rocks in their path. Large amounts of sediment are left behind by glaciers.

Living Things

Plants and animals can also cause weathering. Tree roots can grow in rock cracks. As the roots get bigger, they widen the cracks. Some roots are strong enough to split the rock apart and break it into pieces. When animals dig burrows, or holes, in the ground, they break up rocks on the Earth's surface as well.

Planting in Rocks?

Soil is actually weathered rock particles mixed with humus. Humus is the material left when dead plants and animals break down. Soil has different layers. The top layers have smaller particles due to weathering. Further down, the particles get larger until finally the rock is solid again.

Have you ever seen a large boulder covered by a white or greenish "crust"? These are lichens. Lichens are **organisms** that make their own food. During this process, the lichens release chemicals that break down the rocks or logs the lichens grow on.

Chemicals

Chemicals in the air and in the water are also tough on rocks. Acid rain washes away the surfaces of rocks. It can do great damage to buildings and statues carved out of rocks. Chemicals in groundwater wear down underground rock, such as limestone. These tiny bits of rocks are then carried away by the water.

Lichens

8

Mining

The process of removing minerals from the ground is known as mining. Mining is done at the surface of the Earth, underwater, and underground.

At the Surface

Several different processes have been developed to mine at the surface of the Earth. Placer mining uses a special box to collect sand, gravel, and minerals. When water is washed over the materials, the heavy minerals settle on the bottom of the box.

Open-pit mines use excavators to dig out minerals from large chunks of rock. Explosives are often used to loosen and break up hard rock. The larger rock is then crushed and separated to remove the minerals.

Stripping away the soil and rocks that cover valuable minerals is called *strip mining*. Large shovels remove the land so smaller shovels can dig out the minerals.

A quarry is another type of surface mine. Quarries are areas where large blocks of rock are removed from the ground. Marble, granite, sandstone, limestone, and slate are often mined from rock quarries. Large blocks of stone are cut for buildings and statues. Smaller rocks are dug for use in construction materials, such as cement and bricks. Sand and gravel quarries are also common.

A dredging operation in Idaho

Underwater

Minerals can also be mined from underwater sources. A dredge boat scoops up gravel mixed with minerals from ocean or river bottoms. The minerals are then separated from the gravel, which is returned to the water. Some ships suck up ocean sand that can be sifted through in search of minerals.

Underground

Underground mines are used for digging up minerals and rocks located deep within the Earth. A deep opening, or shaft, for elevators is dug first. Tunnels are dug off the sides of this shaft. People and machinery go down the shaft and into the tunnels. Rocks and minerals are collected from the tunnels and brought back to the surface on the elevators.

Coal is an important mineral mined underground. Coal is burned to supply energy to electric companies. In fact, more than four-fifths of the coal mined in the United States is used to produce electricity.

Mining Ores

An **ore** is a mineral containing valuable metals. These metals are removed from the ore by crushing the mineral and then heating it. This process is called *smelting*. Gold, iron, lead, mercury, aluminum, tin, and copper are just a few of the precious metals found in ores.

A Couple of Ways to Dig Up Coal

Coal found deep beneath the Earth's crust is mined using underground techniques. Coal that is near the surface of the Earth is removed by strip mining.

Gold is often found with quartz growing in a vein, or line, of minerals in rock. Nuggets of gold can be found in sand and gravel beds. Gold is a shiny, heavy metal that is very soft. Often it is mixed with other metals to make it stronger. Gold is a good electrical conductor, so it is used in computers, telephones, air bag triggers, and other equipment that requires reliable electrical signals.

Iron can be found in both hematite and magnetite ore. Iron is the main ingredient in steel, which is used for making strong buildings and heavy machinery.

Lead is mainly found in an ore called *galena*. Lead is a very soft, heavy metal. It is used for making car batteries, windows, and radiation shields used with X-ray equipment.

Galena

Mercury is a metal found in the ore called *cinnabar*. Mercury is a shiny, heavy, poisonous mineral. The silvery liquid in some thermometers is mercury.

Bauxite ore is the main source of aluminum. Aluminum is a very lightweight metal. Packaging (foil) and cans are two common uses of this mineral.

Tin is mined from an ore called *cassiterite*. Tin is often mixed with other elements to form an alloy. Bronze is an alloy of tin and copper. It does not rust and is very strong. Pewter is another alloy. It is a mixture of tin and lead or tin, copper, and antimony. Pewter is often used to make decorative items like jewelry, cups, and figurines.

Copper is a metal that can be found growing in holes in basalt rock. It is also found in the minerals bornite, azurite, and malachite. Copper is good at conducting electricity and heat. It is used to make wiring and cooking pots. Copper mixed with zinc forms the alloy brass.

> **Is Quicksilver a Speedy Kind of Silver?**
>
> No, quicksilver is actually a nickname for mercury. Mercury is a liquid at room temperature. Because it moves so smoothly and easily, it is often called *quicksilver*.

Pewter vase

Copper pot

> **Is It Worth It?**
>
> Getting the metal out of an ore can be costly. The ore must be rich enough (filled with enough metal) to make it worth the time and cost of removing it from the mineral.

9

A World of Natural Wealth

From the beginning of time, rocks and minerals have brought great wealth to the world. Nature provides thousands of wonderful combinations of beauty and practical use.

Throughout history, people have recognized the importance of rocks and minerals. Early people made stone tools for cutting, chopping, and spearing. Rocks and minerals were ground into powders for dyes and paints. Structures such as the Great Pyramids of Egypt and adobe mud houses were made from rocks and minerals. Stones were cut and polished to show wealth and power.

Today, we still use rocks and minerals to make tools, dyes, structures, and jewelry. But over time, we have recognized and found new uses for these elements as well.

The Great Pyramid of Giza in Egypt was built out of two million blocks of stone, each weighing more than two tons (4000 pounds).

A Healthy Diet

Plants and animals (including humans) need minerals to grow and be healthy. If plants don't get enough minerals from the soil, a farmer or gardener can add fertilizer that contains extra minerals. If you don't get enough minerals from the foods you eat, you can take vitamins that have the minerals you need.

> **TRY THIS!**
>
> Check out a vitamin bottle from your house or on a store shelf. What minerals do the vitamins contain?

Fuels

Rocks and minerals are used for energy. All living things contain the element carbon. When organisms die, they are buried and pressed together under layers of rock, soil, and other sediment. Over time, the organisms change into minerals containing carbon. Coal and other **fossil fuels**, such as natural gas and oil, are formed this way.

Oil is the remains of tiny ocean plants and animals. When these organisms sank to the ocean floor, layers of sediment piled on top of them. Over time, heat from the Earth's crust changed the buried remains to oil. Drills dig into rocks and pump the oil to the surface. The oil can then be burned as a fuel.

Building Materials

Concrete is a common building material. It is a combination of gravel, sand, stones, cement, and water. The mineral gypsum is a main ingredient in concrete. The rocks and minerals in concrete have been crushed, mixed, and heated. These rocks are mixed with cement and water to form concrete. When wet concrete dries, it hardens into a tough material that can withstand a lot of pressure.

Clay is another building material. Clay is fine-grained rock particles. When the clay is dried in a kiln (oven), it hardens. The tiny particles fuse together as the water is baked off.

A Mineral of Many Uses

Clay also has a decorative use. It is a main ingredient in ceramic pottery. Throughout history, many cultures have carved or painted pictures on ceramic vases, pots, and bowls. Ceramics continues to be a popular art form today.

Clay can be made into bricks. Bricks fit together tightly to make fireplaces, patios, and buildings. They are held together with a cement paste called *mortar*.

Even the glass in windows is made from rocks and minerals. Sand, limestone, and **soda ash** are the main ingredients of glass. They are mixed, heated to melting, and then poured and blown into different shapes.

Fossils

Rocks and minerals are also teachers. They can provide important information about former life on Earth.

Fossils are found in sedimentary rocks. When animals die, water washes away the soft parts of the body and sediment covers the remains. Over time, the bones absorb minerals from the sediment and become very hard. Shells, teeth, feathers, and eggs can also be fossilized this way.

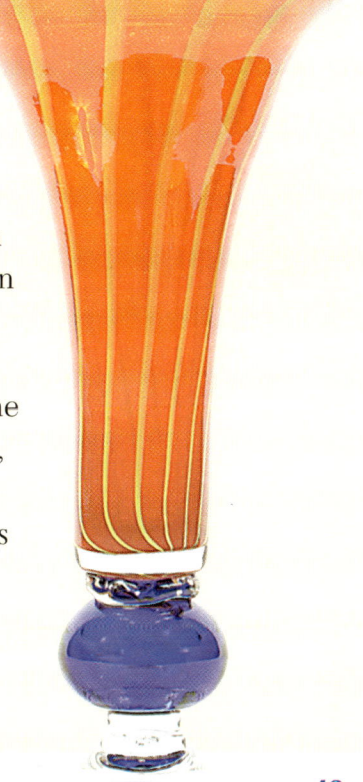

Fossils are a record of past animal life. The order of the layers in fossils provide clues about when the animals lived. Animals found in lower fossil layers lived longer ago than those fossilized in top layers.

Plants can be fossilized too. Leaves and flowers can make prints when the mud around them hardens into rock. Wood can be petrified when minerals replace the living plant, turning it to stone. These rock forms of plants tell scientists about early plant life on Earth.

Rocks from Space

Rocks from space can teach about the universe. Astronauts have collected rocks from planets or moons to study them. They have found that these rocks are very similar to Earth's basalt rocks.

Meteorites are rocks from outer space that land on Earth. They come from asteroids, comets, planets, or the Moon. Meteorites are dark, heavy, and shiny. Many are made of iron and nickel.

❖ ❖ ❖ ❖ ❖

Rocks and minerals are all around us. We walk on them, build with them, use them for fuel, and decorate with them. The Earth itself is layers of solid and liquid rock. So the next time you pick up a rock to skip across the water, stop and appreciate the wealth it holds!

Internet Connections and Related Reading for Rocks and Minerals

http://www.rocksforkids.com/RFK/TableofContents.html
Rock and roll your way through this site that includes information on the formation, identification, collection, and uses of rocks. Try some of the arts and crafts too!

http://www.fi.edu/fellows/fellow1/oct98/index2.html
Become a "rock hound." Learn about the three types of rocks, individual rocks, and rock collecting. Complete a puzzle or quiz to see how much you remember.

http://www.sdnhm.org/kids/minerals
Check out the "Mineral Matters" at this site. Find out how to identify and collect rocks and grow crystals. Play some "mine" games.

http://www.minsocam.org/MSA/K12/K_12.html
Let the Mineralogical Society of America teach you about rock groups, the rock cycle, minerals, and crystals.

http://www.msha.gov/KIDS/KIDSHP.HTM
Explore the Mine Safety and Health Administration's Kid's Page. Find out about what's mined in each state, which minerals every person needs, the history of mining, and mining safety.

http://www.nmnh.si.edu/minsci/images/gallery/gallery.htm
Tour the Smithsonian Rock and Mineral picture gallery to view rocks, minerals, gems, meteorites, and volcanoes.

Dirt: The Scoop on Soil by Natalie M. Rosinsky. Discusses the nature, uses, and importance of soil and the many forms of life that it supports. Picture Window Books, 2003. [IL K–4] (3429606 HB)

The Magic School Bus Inside the Earth by Joanna Cole. A class trip, which includes visiting every layer of the Earth, ends when the class is carried back to school by lava. Scholastic, 1987. [RL 3 IL 1–4] (8954501 PB 8954502 HB)

The Nature and Science of Rocks by Jane Burton and Kim Taylor. An Exploring the Science of Nature book. Gareth Stevens, 1998. [RL 4.9 IL 3–7] (5893106 HB)

Plate Tectonics by Alvin and Virginia Silverstein and Laura Silverstein Nunn. Explains a fundamental concept of science, gives some background, and discusses current applications and developments. Millbrook Press, 1998. [RL 5 IL 5–8] (3112306 HB)

Rocks: Hard, Soft, Smooth, and Rough by Natalie M. Rosinsky. This book discusses the different types of rocks, such as igneous, sedimentary, and metamorphic. Picture Window Books, 2003. [IL K–4] (3429306 HB)

- RL = Reading Level
- IL = Interest Level

Perfection Learning's catalog numbers are included for your ordering convenience. PB indicates paperback. HB indicates hardback.

Glossary

alloy — (AL oy) mixture of metal minerals

atom — (AT uhm) tiny particle that makes up everything in the world

cement — (SIM ent) fine gray powder used to make concrete

conduct — (cuhn DUKT) to allow to pass through; to carry

core — (kor) center of the Earth

crust — (kruhst) thin layer of rock on the top of the Earth

crystal — (KRIS tuhl) solid with atoms that grow in a regular, repeated pattern

deposited — (dee PAH zit ed) dropped or left behind

element — (EL uh ment) nonliving material made up of one type of atom (see separate entry for *atom*)

erosion — (uh ROH zhuhn) movement of rock pieces by wind, water, or ice

evaporate — (ee VAP or ayt) to change from a liquid to a gas

fossil fuel — (FAH suhl fyoul) energy source, such as coal, oil, or natural gas, formed from the remains of dead plants and animals

geology — (jee AHL uh jee) study of the Earth and its features

igneous — (IG nee uhs) rock formed when hot liquid rock from within the Earth is cooled as it approaches or reaches the surface

lava — (LAH vah) liquid rock above the Earth's surface

magma — (MAG mah) liquid rock within the Earth

mantle — (MAN tuhl) layer of the Earth between the crust and the core (see separate entries for *crust* and *core*)

metamorphic	(met uh MOR fik) rock formed when igneous or sedimentary rocks change properties due to heat and/or pressure (see separate entries for *igneous* and *sedimentary*)
mined	(meyend) removed from the ground
mineral	(MIN er uhl) nonliving substance made up of one or more elements (see separate entry for *element*)
ore	(or) mineral in which metal is found
organism	(OR guh niz uhm) living thing
plate	(playt) large piece of the Earth's crust
rock	(rahk) combination of two or more minerals
sediment	(SED uh ment) small pieces of rock, minerals, and soil carried by a body of water
sedimentary	(sed uh MEN tuh ree) rock formed when layers of sediment are pressed together (see separate entry for *sediment*)
soda ash	(SOH duh ash) sodium carbonate; chemical gotten from one of several minerals
spring	(spring) source of water coming from the ground
weathering	(WETH er ing) process of breaking down rocks into smaller pieces

Index

alloys, 25, 40
    brass, 40
    bronze, 40
    pewter, 40
Badlands, 35
clay, 43
concrete, 42–43
crystals, 19–22
Devil's Lake, 35
elements, 5
    periodic table, 5
erosion, 33, 34–36
fossil fuels, 42
fossils, 43–44
geology, 4–5
glass, 43
Grand Canyon, 30, 35
igneous rocks, 27–29, 33
    extrusive, 28–29
    intrusive, 28
layers of the Earth, 7–10
    crust, 9
        plates, 8, 9
    inner core, 7
metals, 8
metamorphic rocks, 31–32, 33
meteorites, 44
minerals, 6, 11–18
    aluminum, 9, 11, 40
    amber, 24
    amphibole, 16
    apatite, 14
    biotite, 16
    calcite, 14, 16, 21, 31
    coal, 11, 39, 42
    copper, 6, 8, 11, 13, 18, 40

minerals (continued)
    corundum, 12, 14
    diamond, 14, 24
    feldspar, 16, 21
    fluorite, 14, 16
    galena, 16, 21, 39
    gold, 8, 12, 13, 15, 18, 21, 39
    graphite, 25
    gypsum, 14, 18, 21, 42
    halite, 16, 21, 26
    hematite, 18, 39
    iron, 7, 8, 39, 44
    jade, 13
    lazurite, 13
    lead, 6, 39
    magnetite, 39
    mercury, 40
    mica, 21
    muscovite, 16
    nickel, 7, 8, 18, 44
    olivine, 21
    orthoclase, 14
    pearl, 24
    pyrite, 15, 21
    pyrophyllite, 18
    quartz, 6, 12, 13, 14, 18, 21, 23, 39
    rutile, 18
    silver, 8, 13, 18, 25
    sulfur, 12, 25
    talc, 13, 14, 21
    tin, 8, 40
    topaz, 14, 21
    turquoise, 21
    zinc, 40
    zircon, 21

mining, 37–40
    underground, 38–39
    underwater, 38
    surface, 37–38
        open-pit, 37
        placer, 37
        quarry, 38
        strip, 38
Mohs, Friedrich, 14
Mohs Scale of Hardness, 14
Mount Rushmore, 28
ores, 39–40
    bauxite, 40
    cassiterite, 40
    cinnabar, 40
properties of minerals, 11–18
    cleavage, 15–17
    color, 12
    habit, 17–18
    hardness, 14–15
    luster, 13
    specific gravity, 18
    streak, 13
rock cycle, 33
rocks, 6, 27–32
    basalt, 29, 44
    gneiss, 32
    granite, 28, 32, 38
    limestone, 30, 31, 32, 38, 43
    marble, 32, 38
    obsidian, 28
    pumice, 29
    quartzite, 32
    sandstone, 30, 32, 38
    shale, 30, 32
    slate, 32, 38
sedimentary rocks, 30, 33
soil, 36
weathering, 33, 34–36